AF812237

PERENNIAL MATH

CONTEST PROBLEMS

2017-2018

Dr. Sylvia Dean was a gifted specialist and administrator for 26 years in Huntsville, Alabama. She created Perennial Math in 2011 upon completion of her doctoral degree. She has hosted many robotics, technology, and mathematics competitions throughout the years and has tirelessly worked on providing educational opportunities to students around the world. It has always been her ambition to leave students with the feeling that "math is cool."

These problems are challenging and students should know up front that scoring a 3 out of 6 on any of these tests is extraordinary and above average. Great problem-solvers learn persistence and appreciation for challenge. Good luck as you encourage your students to take the quest to become great mathletes.

Intermediate

level

Grades 5-6

Name________________________________

PERENNIAL MATH

November 2018 Intermediate Level TEST #1

1) What is the first time after midnight that all the digits on the display of a digital clock are prime numbers?

2) There are 128 tennis players in a tennis tournament. Each player plays only one match per day and all players play the first day. Only winners play the next day until there is only a championship match. If the tournament started on May 30th, on what date will the final match where only two players remaining be played? June ____

3) On the math test Nan received 4 points for correct answers and lost 6 points for incorrect answers. After 17 problems Nan had a score of 8 points. How many problems did she get incorrect?

4) Using the diagram solve for the value of the sum of one turkey, one hat, and one pumpkin.

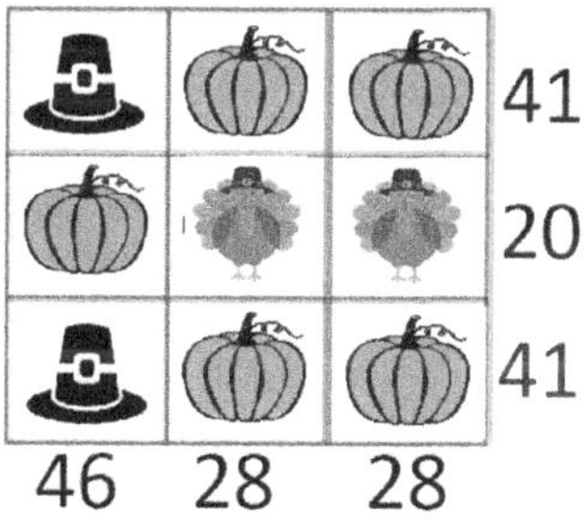

5) A palindrome is a number that reads the same forwards and backwards. How many palindromes are between 100 and 555 and do not contain the digit 6?

6) A wallet has $420 in $5, $10, and $20 bills. The number of $5 bills exceeds twice the number of $10 bills by 4. The number of $20 bills is six fewer than the number of $10 bills. How many $10 bills are there?

PERENNIAL MATH

TEST # 1

November 2018 Intermediate Level Answers

1) What is the first time after midnight that all the digits on the display of a digital clock are prime numbers?	**2:22**
2) There are 128 tennis players in a tennis tournament. Each player plays only one match per day and all players play the first day. Only winners play the next day until there is only a championship match. If the tournament started on May 30th, on what date will the final match where only two players remaining be played? June ____	**5**
3) On the math test Nan received 4 points for correct answers and lost 6 points for incorrect answers. After 17 problems Nan had a score of 8 points. How many problems did she get incorrect?	**6**
4) Using the diagram solve for the value the sum of one turkey, one hat, and one pumpkin. 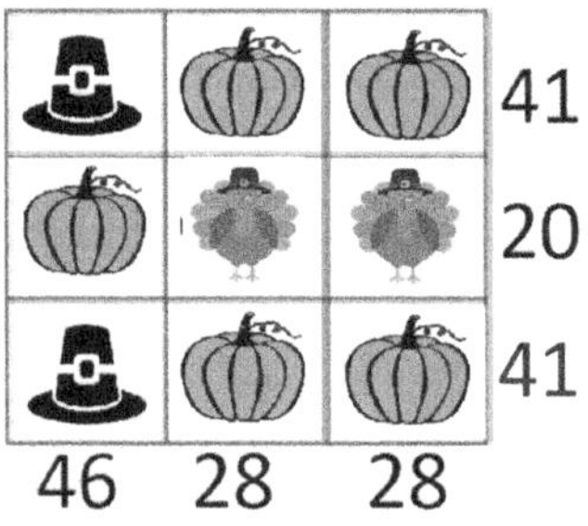	**33**
5) A palindrome is a number that reads the same forwards and backwards. How many palindromes are between 100 and 555 and do not contain the digit 6?	**41**
6) A wallet has \$420 in \$5, \$10, and \$20 bills. The number of \$5 bills exceeds twice the number of \$10 bills by 4. The number of \$20 bills is six fewer than the number of \$10 bills. How many \$10 bills are there?	**13**

PERENNIAL MATH

TEST # 1

November 2018 Intermediate Level Solutions

1) One is not a prime number. The first prime number is 2. The problem did not specify that the numbers had to be different, so the first time after midnight that all the digits on the display of a digital clock are prime numbers is 2:22.

2:22

2) There are 128 tennis players in a tennis tournament. Each player plays only one match per day and all players play the first day. Only winners play the next day until there is only a championship match. So on the first day there were 128 players; on the second day there was 64 players; on the third day there were 32 players; on day four there were 16 players; on the fifth day there were 8 players; on the sixth day there were 4 players; on the seventh day there were 2 players. So seven days beginning on May 30th the final match where only two players remained would be June 5th. The answer is June _5_.

5

3) On the math test Nan received 4 points for correct answers and lost 6 points for incorrect answers. After 17 problems Nan had a score of 8 points. Make a chart to find how many problems she got incorrect. If you start in the middle with 9 correct and 8 incorrect, you will see that for each correct problem you add, your total points actually increases by 10. Nan had 6 problems incorrect.

Correct +4	9	10	11
Incorrect -6	8	7	6
Points for Correct	36	40	44
Points for Incorrect	-48	-42	-36
Total Points	-12	-2	8

6

PERENNIAL MATH

TEST #1

November 2018 Intermediate Level Solutions

4) Use algebra. Let H = hat, P = pumpkin, and T = turkey. Add the equations together so that you have 2 of each and divide by 2 so that you find how the value of H + P + T.

$$\begin{array}{rl} & 2H + P = 46 \\ + & P + 2T = 20 \\ \hline & 2H + 2P + 2T = 66 \end{array}$$

Now divide both sides in half. H + P + T = 33.

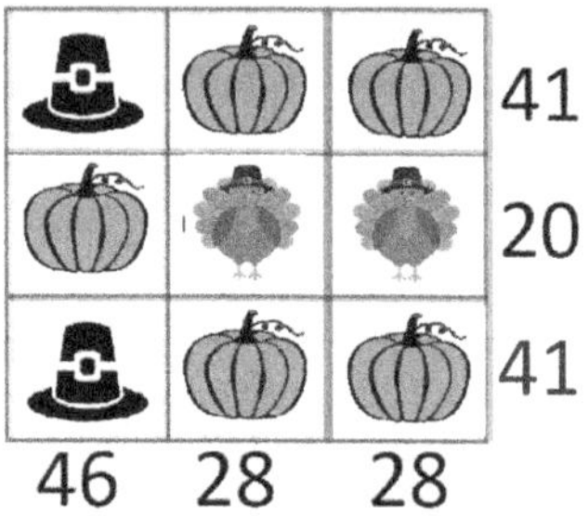

33

5) Here is a list of palindromes between 100 and 555 with the numbers containing the digit 6 marked out. 101, 111, 121, 131, 141, 151, ~~161,~~ 171, 181, 191, 202, 212, 222, 232, 242, 252, ~~262,~~ 272, 282, 292, 303, 313, 323, 333, 343, 353, ~~363,~~ 373, 383, 393, 404, 414, 424, 434, 444, 454, ~~464,~~ 474, 484, 494, 505, 515, 525, 535, 545. The problem did not state that 555 was inclusive, so do not count it. There are 41 palindromes.

41

6) A wallet has \$420 in \$5, \$10, and \$20 bills. The number of \$5 bills exceeds twice the number of \$10 bills by 4. The number of \$20 bills is six fewer than the number of \$10 bills. Make a chart to find the number of \$10 bills. Start with seven \$10 bills because there are six fewer \$20 bills. You will notice that each time the number of \$10 bills increases, the sum increases by \$40. There are thirteen \$10 bills.

$5 Bills	$10 Bills	$20 Bills	Sum
18	7	1	$180
20	8	2	$220
22	9	3	$260
24	10	4	$300
26	11	5	$340
28	12	6	$380
30	13	7	$420

13

Name________________________________

PERENNIAL MATH

TEST #2

December 2018 Intermediate Level

1) How many prime numbers are between 13 and 113?	
2) The numbers 11 through 16 were put in a hat and Bob drew out a number. If the units digit of the sum of the remaining numbers is 7, what number did Bob draw out?	
3) How many paths exist from bottom corner X to top corner Y? You may only move to the right or up or diagonally upward. 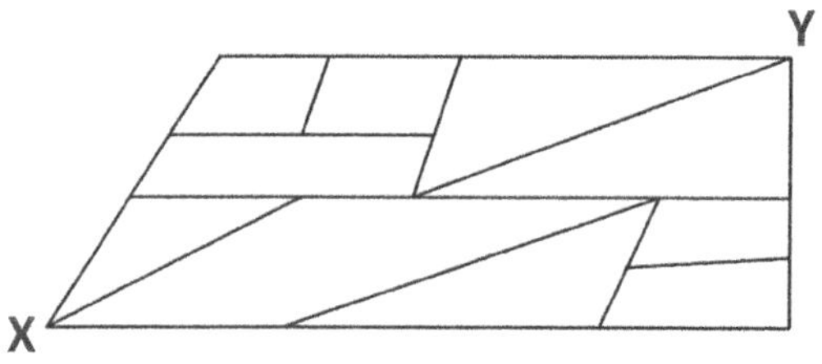	
4) In the land of Zuphronx they use three coin types: amixls, debixls, and remixls. Four amixls and one debixl equals one remixl. Two debixls and one remixl equals 10 amixls. How many amixls would equal one remixl?	
5) A bag has 10 yellow balls and 5 green balls. Two balls are taken from the bag without replacement. What is the probability both balls are yellow? (Answer as a simplified fraction)	
6) In Bob's Gym there are six more 50-pound weights than 100-pound weights and three times as many 20-pound weights as 50-pound weights, for a total of 3180 pounds. How many 20-pound weights are there?	

PERENNIAL MATH TEST #2

December 2018 Intermediate Level Answers

1) How many prime numbers are between 13 and 113?	23
2) The numbers 11 through 16 were put in a hat and Bob drew out a number. If the units digit of the sum of the remaining numbers is 7, what number did Bob draw out?	14
3) How many paths exist from bottom corner X to top corner Y? You may only move to the right or up or diagonally upward. 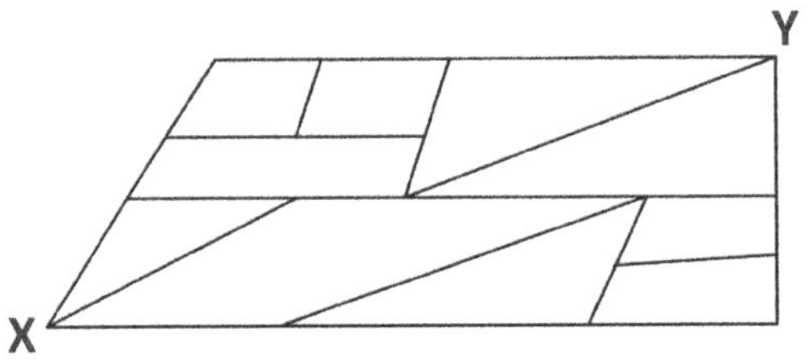	13
4) In the land of Zuphronx they use three coin types: amixls, debixls, and remixls. Four amixls and one debixl equals one remixl. Two debixls and one remixl equals 10 amixls. How many amixls would equal one remixl?	6
5) A bag has 10 yellow balls and 5 green balls. Two balls are taken from the bag without replacement. What is the probability both balls are yellow? (Answer as a simplified fraction)	3/7
6) In Bob's Gym there are six more 50-pound weights than 100-pound weights and three times as many 20-pound weights as 50-pound weights, for a total of 3180 pounds. How many 20-pound weights are there?	54

PERENNIAL MATH

TEST #2

December 2018 Intermediate Level Solutions

1) Here are the prime numbers between 13 and 113: ~~2, 3, 5, 7, 11, 13,~~ 17, 19, 23, 29, 31, 37, 41, 43, 47, 53, 59, 61, 67, 71, 73, 79, 83, 89, 97, 101, 103, 107, 109, ~~113~~. There are twenty-two prime numbers between 13 and 113.

23

2) The numbers 11 through 16 were put in a hat and Bob drew out a number. The units digit of the sum of the remaining numbers is 7. Act this out to find what number Bob drew out.

Remove the 16: 11 + 12 + 13 + 14 + 15 = 65
Remove the 15: 11 + 12 + 13 + 14 + 16 = 66
Remove the 14: 11 + 12 + 13 + 15 + 16 = 67

Bob drew out the number 14.

14

3) There are 13 paths from X to Y that go up, right or diagonal.

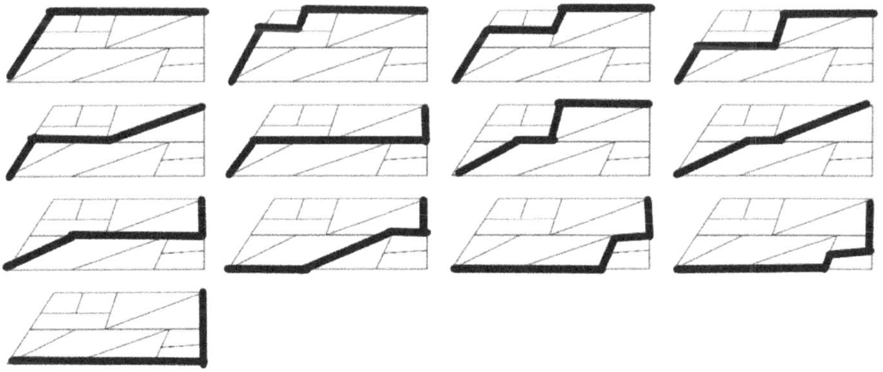

13

PERENNIAL MATH

TEST # 2

December 2018 Intermediate Level Solutions

4) Use algebra to find how many amixls are equal to one remixl.
Since 4 amixls and 1 debixl equals 1 remixl, then 2 debixls plus 4 amixls and 1 debixl equal 10 amixls. If 3 debixls and 4 amixls equal 10 amixls, then 1 debixl equals 2 amixls. Now 4 amixls plus 2 amixls equal 1 remixl or 6 amixls.

$4A + 1D = 1R$ {equation 1}

$2D + 1R = 10A$ {substitute $(1R = 4A + 1D)$}

$2D + 4A + 1D = 10A$

$3D = 6A$

$D = 2A$

$4A + 1D = 1R$ {now substitute $(D = 2A)$ into equation 1}

$6A = 1R$

6

5) A bag has 10 yellow balls and 5 green balls. Two balls are taken from the bag without replacement. What is the probability both balls are yellow? There are 15 total balls. The probability of getting a yellow ball on the first draw is 10/15. After the first ball is pulled out, then the probability of getting a yellow ball is 9/14. To find the probability of getting two yellow balls without replacement multiply:

$$\frac{10}{15} \times \frac{9}{14} = \frac{90}{210} = \frac{3}{7}$$

3/7

6) In Bob's Gym there are six more 50-pound weights than 100-pound weights and three times as many 20-pound weights as 50-pound weights, for a total of 3180 pounds. Make a chart to find the number of 20-pound weights. Start with one 100-pound weight. Notice that each time you add a 100-pound weight that the total weight increases by 210 pounds. There are 54 20-pound weights.

20 lb.	50 lb.	100 lb.	Total Weight
21	7	1	870
24	8	2	1,080
27	9	3	1,290
30	10	4	1,500
33	11	5	1,710
36	12	6	1,920
39	13	7	2,130
42	14	8	2,340
45	15	9	2,550
48	16	10	2,760
51	17	11	2,970
54	18	12	3,180

54

Name________________________________

PERENNIAL MATH TEST #3

January 2019 Intermediate Level

1) What two-digit number is two times the sum of its digits?

2) Tracey writes a sequence of odd numbers as follows:
Row 1: 13579
Row 2: 135791133557799
Row 3: 135791133557799111333555777999.
If she continues this pattern in subsequent rows, what will be the 100th number that she writes if you count from the very beginning of the first row?

3) The ratio of children to adults at the magic show is 3:1. There are a total of 484 people there. How many are children?

4) How many triangles are in this diagram?

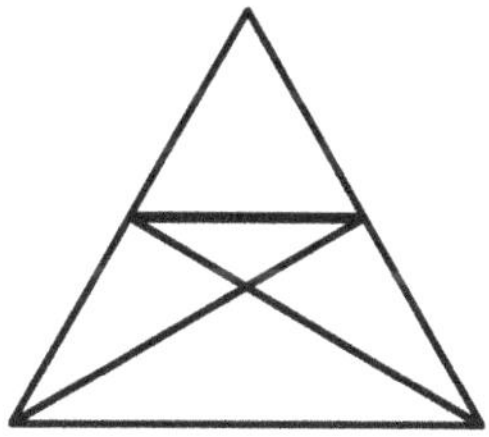

5) Phil sold pizza at the carnival for $1.75 a slice or $9.50 for a whole pizza. He sold 76 slices of pizza and 64 whole pizzas. He gave 15% of his money to the carnival fund. How much money did he have left?

6) If Andrew gives Gene 10 gumdrops, they both will have the same amount. However, if Gene gives Andrew 10 gumdrops then Andrew will have 5 times as many gumdrops as him. How many gumdrops does Andrew have?

TEST #3

PERENNIAL MATH

January 2019 Intermediate Level Answers

1) What two-digit number is two times the sum of its digits?	18
2) Tracey writes a sequence of odd numbers as follows: Row 1: 13579 Row 2: 135791133557799 Row 3: 135791133557799111333555777999. If she continues this pattern in subsequent rows, what will be the 100th number that she writes if you count from the very beginning of the first row?	9
3) The ratio of children to adults at the magic show is 3:1. There are a total of 484 people there. How many are children?	363
4) How many triangles are in this diagram? 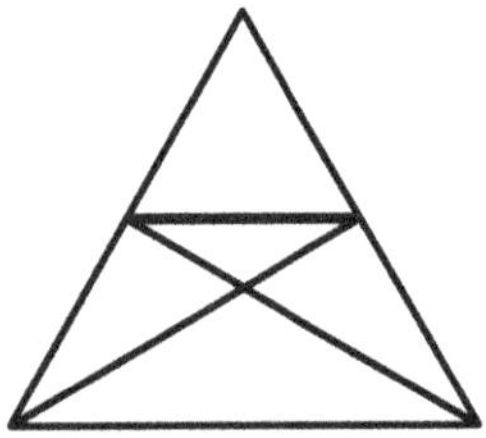	12
5) Phil sold pizza at the carnival for $1.75 a slice or $9.50 for a whole pizza. He sold 76 slices of pizza and 64 whole pizzas. He gave 15% of his money to the carnival fund. How much money did he have left?	629.85
6) If Andrew gives Gene 10 gumdrops, they both will have the same amount. However, if Gene gives Andrew 10 gumdrops then Andrew will have 5 times as many gumdrops as him. How many gumdrops does Andrew have?	40

PERENNIAL MATH

TEST # 3

January 2019 Intermediate Level Solutions

1) What two-digit number is two times the sum of its digits? The number 18 is equal to 2 (1+8).

18

2) Tracey writes the odd number 13579 in the first row. She writes two of them 135791133557799 in the next row. Then she adds three of each number to the list in the third row 135791133557799111333555777999. Act this out to find the 100th number.

13579 = 5 numbers

135791133557799 = 15 numbers

135791133557799111333555777999 = 30

Row four will have 30 numbers plus an additional 5(4) or 20 numbers. That will be a total of 5 + 15 + 30 + 50 = 100 numbers. The last number written in the fourth row was 9 which was the 100th number counting from the beginning.

9

3) The ratio of children to adults at the magic show is 3:1. There are a total of 484 people there. Since there are 3 groups of children to one group of adults, there are 4 portions that total 484. Divide 484 by 4 to get the amount of one portion. 484/4 = 121. Since there were three groups or portions of children, then there are 3 times 121 = 363 children.

363

PERENNIAL MATH TEST # 3

January 2019 Intermediate Level Solutions

4) Count the triangles systematically. There are 5 individual small triangles. Next there are 4 triangles comprised of two triangles. There are also 2 triangles comprised of three triangles. Finally there is one triangle made of all five triangles. There are 12 total triangles.

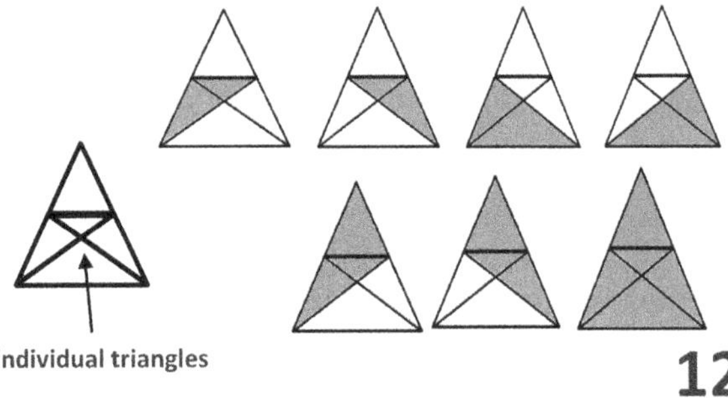

12

5) Phil sold pizza at the carnival for $1.75 a slice or $9.50 for a whole pizza. He sold 76 slices of pizza at $1.75 each. 76 x $1.75 = $133. He sold 64 whole pizzas for $9.50. 64 x $9.50 = $608. $133 + $608 = $741. $741 x 15% = $741 x .15 = $111.15. Subtract to find out how much money he had left. $741 - $111.15 = $629.85.

629.85

6) Use algebra to find how many gumdrops Andrew has. Let A represent Andrew and G represent Gene. If Andrew gives Gene 10 gumdrops, they both will have the same amount: A - 10 = G + 10.

A = (G+20) {equation 1}

However, if Gene gives Andrew 10 gumdrops then Andrew will have 5 times as many gumdrops as him: (G - 10)5 = A + 10

5G - 50 = A + 10 {equation 2}

Now substitute the first equation (G + 20) in for A in the second equation.

5G - 50 = (G + 20) + 10

5G - 50 = G + 30

4G = 80

G = 20

Substitute 20 for G into the first equation A = (G + 20).

A = 20 + 20 = 40

Therefore Andrew has 40 gumdrops, Gene has 20 and their total is 60.

40

Name________________________________

PERENNIAL MATH

TEST #4

February 2019 Intermediate Level

1) 5 + 3 x 4 = ____ + 2	
2) Brian and Mark both have 48 cards. Brian gave Mark half his cards. Mark then turned around and gave Brian one fourth of the cards he had now. After that Mark had ____ more cards than Brian.	
3) If A # B = 3B + 2A + 1, then what is the value of 5 # (6 # 2)?	
4) Find the area of this figure that has a square cut out of it. 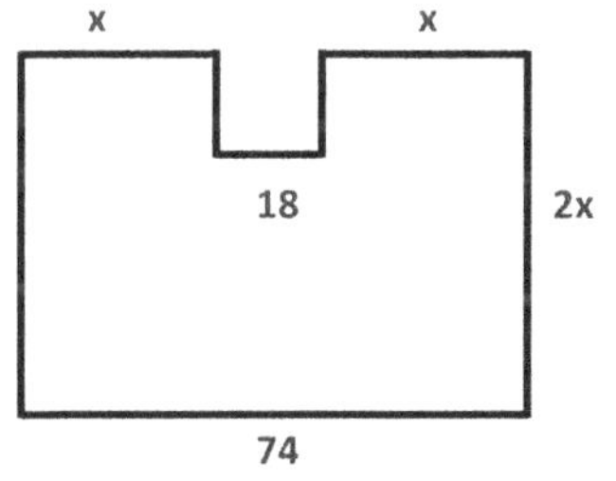	
5) Heather and Suzanne have 20 suckers in total. If Heather gives Suzanne 3 suckers, then she will have 3 times as much as Suzanne. How many suckers does Heather have originally?	
6) John made a list of three-digit whole numbers, and every digit he used was even. At most how many different numbers were on his list?	

PERENNIAL MATH TEST #4

February 2019 Intermediate Level Answers

1) 5 + 3 x 4 = ____ + 2

15

2) Brian and Mark both have 48 cards. Brian gave Mark half his cards. Mark then turned around and gave Brian one fourth of the cards he had now. After that Mark had ____ more cards than Brian.

12

3) If A # B = 3B + 2A + 1, then what is the value of 5 # (6 # 2)?

68

4) Find the area of this figure that has a square cut out of it.

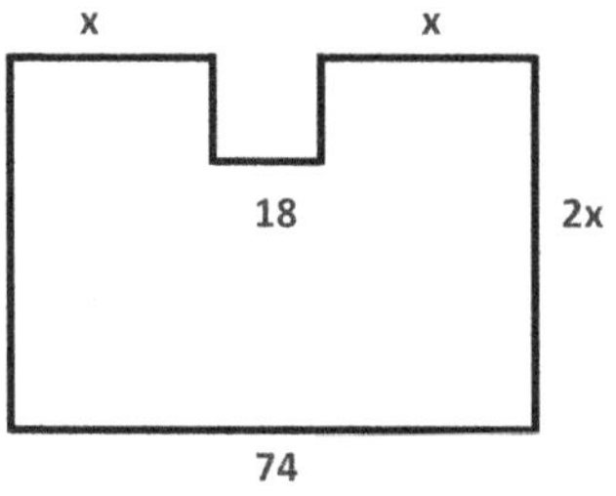

3820

5) Heather and Suzanne have 20 suckers in total. If Heather gives Suzanne 3 suckers, then she will have 3 times as much as Suzanne. How many suckers does Heather have originally?

18

6) John made a list of three-digit whole numbers, and every digit he used was even. At most how many different numbers were on his list?

100

PERENNIAL MATH

TEST #4

February 2019 Intermediate Level Solutions

1) 5 + 3 x 4 = ____ + 2
First use the order of operations to solve for the value of the left side of the equations 5 + (3 x 4) = 17. Now subtract 2 from both sides. The missing number is 15 because 15 + 2 = 17 which is equivalent to the first equation.

15

2) Brian and Mark both have 48 cards. Brian gave Mark half his cards. Mark then turned around and gave Brian one fourth of the cards he had now to Brian. After that Mark had ____ more cards than Brian. Act this out to solve for the final step.

Brian	Mark	Transaction	Brian New	Mark New
48	48	Brian gave Mark half his cards (24)	24	72
24	72	Mark gave Brian one fourth his cards (18)	42	54
42	54	Mark - Brian = (54 - 42) = 12		

Mark had 12 more cards than Brian in the end.

12

3) If A # B = 3B + 2A + 1, then what is the value of 5 # (6 # 2)?
First solve for the equation in the parenthesis (6 # 2).
(6 # 2) = 3(2) + 2(6) + 1
= 6 + 12 + 1
= 19
Now solve for 5 # 19
5 # 19 = 3(19) + 2(5) + 1
= 57 + 10 + 1
= 68

68

PERENNIAL MATH TEST #4

February 2019 Intermediate Level Solutions

4) To find the area of the figure, first solve for x. Subtract 18 to find what the remaining amount would be for each x on the top of the rectangle. 74 – 18 = 56. 2x = 56. x = 56/2 = 28. The area of the shape before the square is cut out is 74 x 56 = 4144 square units. Now subtract out the area of the square which is 18^2 = 324 square units. 4144 – 324 = 3820 square units.

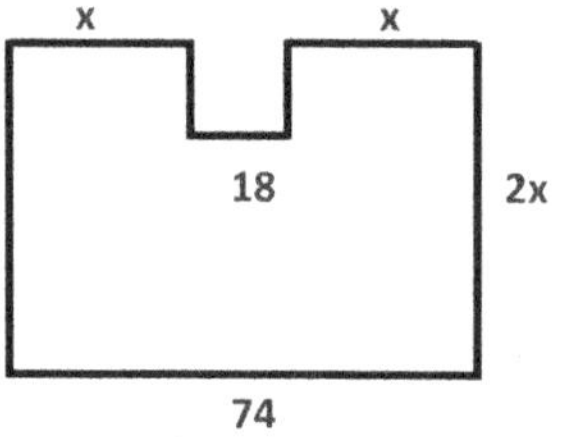

3820

5) Heather and Suzanne have 20 suckers in total. If Heather gives Suzanne 3 suckers, then she will have 3 times as much as Suzanne. Make a table to find how many suckers Heather had originally. Start with 10 suckers a piece and increase the number that Heather has until it is 3 times as much as Suzanne. Heather had 18 suckers originally.

Heather	Suzanne	H - 3	S + 3	Total Suckers
10	10	7	13	20
11	9	8	12	20
12	8	9	11	20
13	7	10	10	20
14	6	11	9	20
15	5	12	8	20
16	4	13	7	20
17	3	14	6	20
18	2	15	5	20

18

6) John made a list of three-digit whole numbers, and every digit he used was even. At most how many different numbers were on his list?
For the ones digit, the digits could be 0, 2, 4, 6, and 8.
For the tens digit, the digits could also be 0, 2, 4, 6, and 8.
For the hundreds digit, the digits could be 2, 4, 6, 8 (Note you cannot have a 0 in the hundreds digit). So since repetition of a number is possible, then there are
4 (hundreds digits) x 5 (tens digits) x 5 (ones digits) = 100 possible three-digit numbers where every digit is an even number.

100

Name________________________________

PERENNIAL MATH

TEST #1

January 2019 Intermediate Level

1) 15 + 35 + 42 = 23 x ___	
2) A car got 26 miles per gallon using gasoline that cost $2.75 per gallon. What was the cost of the gasoline used in driving the car 416 miles?	
3) How many triangles of any size are in this figure? 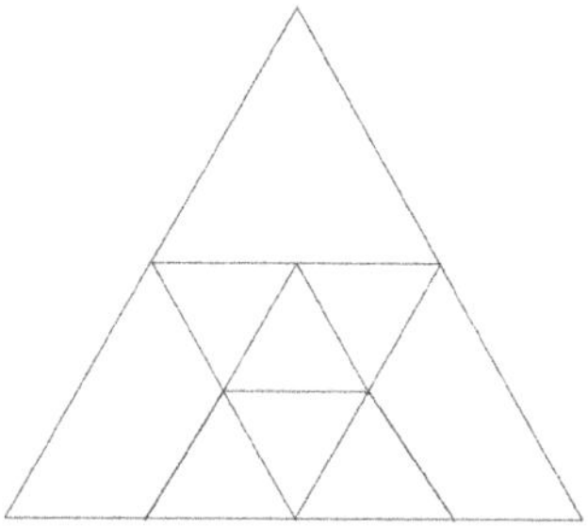	
4) Barb is canning jars of salsa. Each jar requires 2/3 pound of tomatoes. She needs to can 25 jars. She buys 2 pound trays of tomatoes at the market. How many trays of tomatoes must she buy to can the 25 jars of salsa?	
5) Marcia has 28 bracelets. 16 of the bracelets have beads, 15 of the bracelets have charms, and 4 of the bracelets have neither beads nor charms. How many of the bracelets have both beads and charms on them?	
6) The sum of five consecutive odd integers is 2025. What is the sum of the digits in the five consecutive integers?	

PERENNIAL MATH TEST # 1

January 2019 Intermediate Level Answers

1) 15 + 35 + 42 = 23 x ___	4
2) A car got 26 miles per gallon using gasoline that cost $2.75 per gallon. What was the cost of the gasoline used in driving the car 416 miles?	44
3) How many triangles of any size are in this figure? 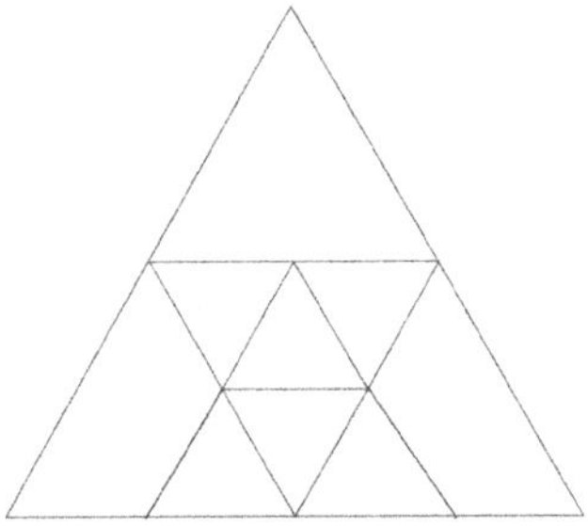	12
4) Barb is canning jars of salsa. Each jar requires 2/3 pound of tomatoes. She needs to can 25 jars. She buys 2 pound trays of tomatoes at the market. How many trays of tomatoes must she buy to can the 25 jars of salsa?	9
5) Marcia has 28 bracelets. 16 of the bracelets have beads, 15 of the bracelets have charms, and 4 of the bracelets have neither beads nor charms. How many of the bracelets have both beads and charms on them?	7
6) The sum of five consecutive odd integers is 2025. What is the sum of the digits in the five consecutive integers?	45

PERENNIAL MATH TEST #1

January 2019 Intermediate Level Solutions

1) 15 + 35 + 42 = 23 x ___
92 = 23 x ___
92/23 = 4

4

2) A car got 26 miles per gallon using gasoline that cost $2.75 per gallon. Approximately what was the cost, in dollars, of the gasoline used in driving the car 416 miles?
Divide 416 by 26 to see how many gallons of gas were used, 416/26 = 16. The cost of 16 gallons of gas is 16 x $2.75 = $44.

44

3) There are 12 triangles in the figure.

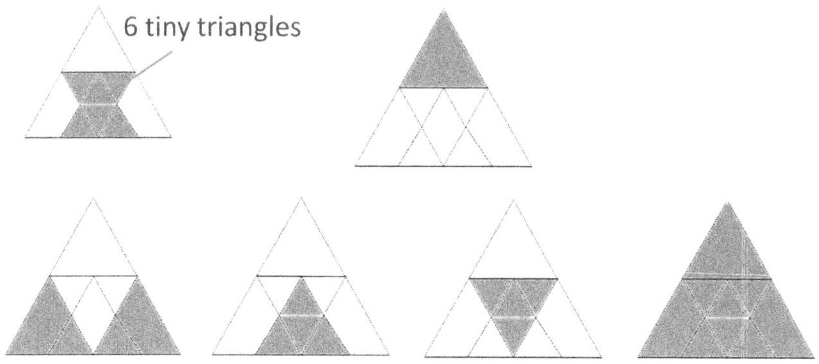

12

PERENNIAL MATH

TEST #1

January 2019 Intermediate Level Solutions

4) Barb is canning jars of salsa. Each jar requires 2/3 pound of tomatoes. She needs to can 25 jars. She buys 2 pound trays of tomatoes at the market. To find how many trays of tomatoes she must buy to can the 25 jars of salsa, first find the number of pounds of tomatoes she is going to need. $25 \times 2/3 = 50/3 = 16\ 2/3$. Since the trays of tomatoes come in 2 pounds, eight trays will not be enough. She must buy 9 trays.

9

5) Marcia has 28 bracelets. 16 of the bracelets have beads, 15 of the bracelets have charms, and 4 of the bracelets have neither beads nor charms. How many of the bracelets have both beads and charms on them? Use a Venn Diagram. The number of bracelets with beads only = 16 - x. The number of bracelets with charms only = 15 - x. Now, to solve for the middle section use algebra.

$$28 = (16 - x) + (15 - x) + x + 4$$
$$28 = 35 - x$$
$$x = 7$$

There are 7 bracelets with both beads and charms.

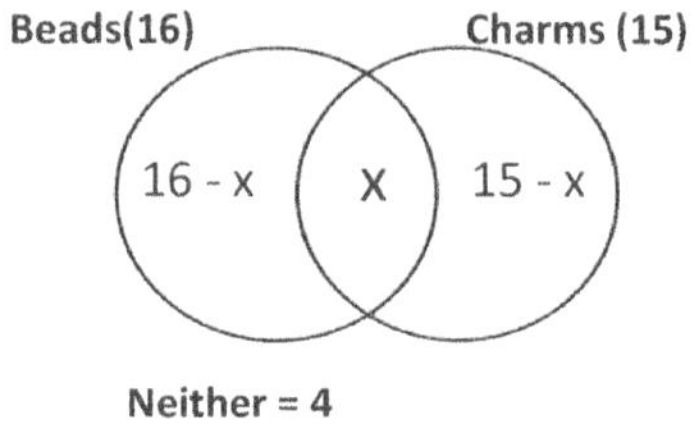

7

6) The sum of five consecutive odd integers is 2025. What is the sum of the digits in the five consecutive integers? First solve for the five odd integers by finding their average. 2025/5 = 405. That would make 405 the middle number of the five integers. The five integers are 401, 403, 405, 407, and 409. The sum of all their digits is:

$$4 + 1 + 4 + 3 + 4 + 5 + 4 + 7 + 4 + 9 =$$
$$4(5) + (1 + 3 + 5 + 7 + 9) =$$
$$20 + 25 = 45$$

45

Name________________________________

PERENNIAL MATH

TEST #2

February 2019 Intermediate Level

1) 420 minutes before 3 p.m. is _____? (include a.m. or p.m. in your answer)	
2) How many 2 by 2 rectangles can fit into a 20 by 18 rectangle?	
3) The diagram below shows a starting number and ending number when "→" is applied. What is the result after the operation "→" is applied to 690? 35 → 7 82 → 9 50 → 4 397 → 18 690 → ?	
4) Jeff failed 7 of his first 11 math quizzes. He passed 75% of his remaining quizzes which made him pass 2/3 of the overall quizzes for the semester. How many total quizzes did he pass this semester?	
5) Twelve boys were comparing their 57 marbles. Alan has more marbles than each of the other boys. Every boy has a prime number of marbles including Alan. What is the most number of marbles that Alan could have?	
6) The doughnut machine packages 1 doughnut every three seconds. It takes 160 doughnuts to complete a full crate. At this rate, how many crates will the doughnut machine fill in 4 hours?	

PERENNIAL MATH

TEST #2

February 2019 Intermediate Level Answers

1) 420 minutes before 3 p.m. is _____? (include a.m. or p.m. in your answer)	8 a.m.
2) How many 2 by 2 rectangles can fit into a 20 by 18 rectangle?	90
3) The diagram below shows a starting number and ending number when "→" is applied. What is the result after the operation "→" is applied to 690? 35 → 7 82 → 9 50 → 4 397 → 18 690 → ?	14
4) Jeff failed 7 of his first 11 math quizzes. He passed 75% of his remaining quizzes which made him pass 2/3 of the overall quizzes for the semester. How many total quizzes did he pass this semester?	34
5) Twelve boys were comparing their 57 marbles. Alan has more marbles than each of the other boys. Every boy has a prime number of marbles including Alan. What is the most number of marbles that Alan could have?	31
6) The doughnut machine packages 1 doughnut every three seconds. It takes 160 doughnuts to complete a full crate. At this rate, how many crates will the doughnut machine fill in 4 hours?	30

PERENNIAL MATH TEST #2

February 2019 Intermediate Level Solutions

1) 420 minutes before 3 p.m. is ______? (include a.m. or p.m. in your answer)
420 minutes/60 = 7 hours. Seven hours before 3 p.m. is 8 a.m.

8 a.m.

2) How many 2 by 2 rectangles can fit into a 20 by 18 rectangle? First find the area of the larger rectangle and then divide it by the area of the smaller rectangle. The area of the larger rectangle is 20 x 18 = 360 units^2. The area of the small rectangle is 2 x 2 = 4 units^2. 360/4 = 90. You can fit 90 smaller rectangles in the larger rectangle.

90

3) The diagram below shows a starting number and ending number when "→" is applied. What is the result after the operation "→" is applied to 690?

	Sum of digits	Sum of digits - 1
35 → 7	8 →	7
82 → 9	10 →	9
50 → 4	5 →	4
397 → 18	19 →	18
690 → ?	15 →	?

The pattern consists of the numbers on the right is the sum of the digits on the left minus 1. Since the sum of the digits of 690 is 15, then the number on the right must be 15 - 1 = 14.

14

PERENNIAL MATH

TEST # 2

February 2019 Intermediate Level Solutions

4) Jeff failed 7 of his first 11 math quizzes. He passed 75% of his remaining quizzes which made him pass 2/3 of the overall quizzes for the semester. Make a chart to find the total quizzes he pass this semester. First consider how many quizzes he passed. He only passed 4/11 of the first quizzes. Then he averaged 75% on the remaining quizzes which is 3 out of 4. So, add 3/4 to the chart until you reach a total amount that is 2/3. He passed 34 out of 51 quizzes which is equal to 2/3.

Quizzes Passed	4	7	10	13	16	19	22	25	28	31	34
Quizzes Taken	11	15	19	23	27	31	35	39	43	47	51
Percentage		46.7	52.6	56.5	59.2	61.2	62.8	64.1	65.1	65.9	66.7

34

5) Twelve boys were comparing their 57 marbles. Alan has more marbles than each of the other boys. Every boy has a prime number of marbles including Alan. To find the most amount of marbles Alan can have, make a list of the prime numbers and see which twelve have a sum of 57. Then select the largest prime number.

2, 3, 5, 7, 11, 13, 17, 19, 23, 29, 31, 37, 41, 43, 47, 53

Since the problem did not state that the other boys had to have different numbers of marbles, then the other 11 boys could have 2 marbles. This would give Alan 57 - (11)(2) = 35. This is not a prime number, but the closest lower prime number is 31. Of the other eleven boys, 7 boys could have 2 marbles and the other 4 boys could have 3 marbles. 31 + (7 x 2) + (4 x 3) = 57. So the most marbles Alan could have is 31 marbles.

31

6) The doughnut machine packages 1 doughnut every three seconds. It takes 160 doughnuts to complete a full crate. To find how many crates the doughnut machine will fill in 4 hours, convert the packaging from seconds to hours and divide by the doughnuts per crate.

1 doughnut every 3 seconds = 20 doughnuts a minute
20 doughnuts per minute x 60 minutes = 1200 doughnuts per hour
1200 doughnuts x 4 hours = 4800 doughnuts
4800 doughnuts ÷ 160 doughnuts per crate = 30 crates.

30

Name________________________________

PERENNIAL MATH

TEST #3

March 2019 Intermediate Level

1) 1, 2, 6, 24, _____, _____ What would be the sum of the two missing numbers that complete this sequence?

2) Rachel bought a bracelet for $120, sold it for $130, bought it back for $140 and then sold it for $150. How much money did she make?

3) Jack is three times as old as Jill. In 6 years Jack will be twice as old as Jill. How old is Jack now?

4) What is the sum of one star, one circle and one hexagon?

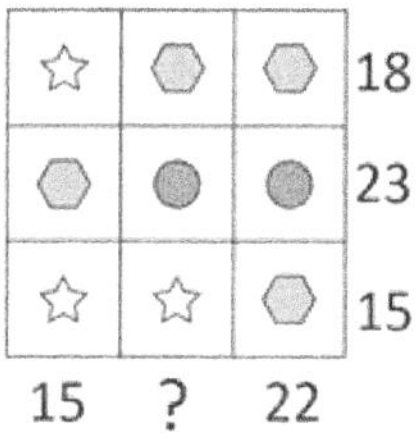

5) If 2/3 cup of kitten chow will feed 8 kittens, how many kittens will 4 cups of kitten chow feed assuming they eat at the same rate?

6) A class of 50 students completed a survey about food. 4 students liked pizza and tacos, but not hamburgers. 12 students liked hamburgers and tacos, but not pizza. 10 students liked pizza and tacos. 18 students liked pizza and hamburgers. Altogether, 30 students liked pizza, 32 students liked hamburgers, and 28 students liked tacos. If all students like something, how many students like tacos only?

PERENNIAL MATH

TEST #3

March 2019 Intermediate Level Answers

1) 1, 2, 6, 24, ____, _____ What would be the sum of the two missing numbers that complete this sequence?

840

2) Rachel bought a bracelet for $120, sold it for $130, bought it back for $140 and then sold it for $150. How much money did she make?

20

3) Jack is three times as old as Jill. In 6 years Jack will be twice as old as Jill. How old is Jack now?

18

4) What is the sum of one star, one circle and one hexagon?

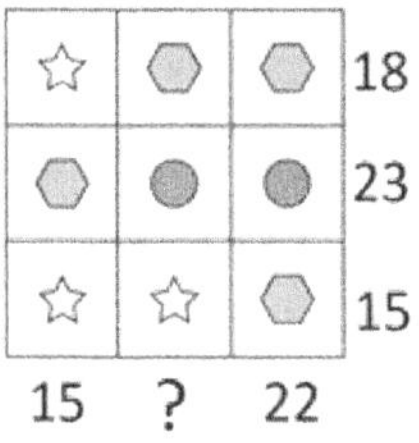

19

5) If 2/3 cup of kitten chow will feed 8 kittens, how many kittens will 4 cups of kitten chow feed assuming they eat at the same rate?

48

6) A class of 50 students completed a survey about food. 4 students liked pizza and tacos, but not hamburgers. 12 students liked hamburgers and tacos, but not pizza. 10 students liked pizza and tacos. 18 students liked pizza and hamburgers. Altogether, 30 students liked pizza, 32 students liked hamburgers, and 28 students liked tacos. If all students like something, how many students like tacos only?

6

PERENNIAL MATH

TEST # 3

March 2019 Intermediate Level Solutions

1) 1, 2, 6, 24, _____, ______ What would be the sum of the two missing numbers that complete this sequence? If n represents the place the item is in the sequence, then the numbers increase by multiplying the previous number by n. For instance the number 24 is the fourth item and it is 4 times the previous number 6. The fifth item will be 24 by 5. 24 x 5 = 120. Then the sixth item is 120 x 6 = 720. Therefore, the sum of the two missing numbers is 120 +720 = 840.

840

2) Act this out: Pretend she has $200. She buys a bracelet for $120 leaving her with $80. She sells the bracelet for $130 which makes a total of $210 back in her wallet. She buys it back for $140 leaving her with $70. Finally she sells it for $150 and that will give her with $220 back in her wallet. Since she started out with $200, that means she made a profit of $20.

20

3) Jack is three times as old as Jill. In 6 years Jack will be twice as old as Jill. Make a chart to see how old Jack is now. Start with Jill as 1 year old and increase her age until Jack is twice as old as Jill in six years. So, Jack is currently 18 years old.

Jill	Jack	Jill + 6	Jack + 6
1	3	7	9
2	6	8	12
3	9	9	15
4	12	10	18
5	15	11	21
6	18	12	24

18

PERENNIAL MATH

TEST # 3

March 2019 Intermediate Level Solutions

4) Use algebra to find the sum of one star, one circle and one hexagon. Let S = star, C = circle, and H = hexagon. Add up the equations that give you 2 of each item and then divide by 2.

$$2C + H = 23 \quad \text{\{middle row\}}$$
$$+ \quad H + 2S = 15 \quad \text{\{first column\}}$$
$$2C + 2H + 2S = 38 \quad \text{\{Divide both sides by 2\}}$$
$$C + H + S = 19$$

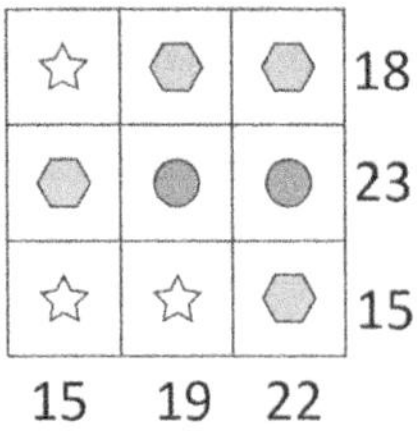

19

5) If 2/3 cup of kitten chow will feed 8 kittens, how many kittens will 4 cups of kitten chow feed assuming they eat at the same rate? Triple the amount of kitten chow. 2/3 cups x 3 = 2 cups. If 2/3 cups feeds 8 kittens and we triple that amount it will feed triple the amount of kittens which is 8 x 3 = 24 kittens. Therefore if 2 cups of kitten chow feeds 24 kittens, then we can assume that double the amount of kitten chow (4 cups) will feed double the amount of kittens. 4 cups will feed 48 kittens.

48

6) Use a Venn Diagram. All regions total 50 students. Label the totals for each category: 30 students liked pizza, 32 liked hamburgers, and 28 liked tacos. Fill in these areas: 4 students liked pizza and tacos but not hamburgers; 12 students liked hamburgers and tacos, but not pizza. Use these to solve for missing areas. 10 students liked pizza and tacos [Region V + 4], Region V = 10 - 4 = 6. 18 students liked pizza and hamburgers [Region IV + Region V], so Region IV = 18 - 6 = 12. Finally, solve for Region II, 28 - 4 - 6 - 12 = 6 students who like tacos only.

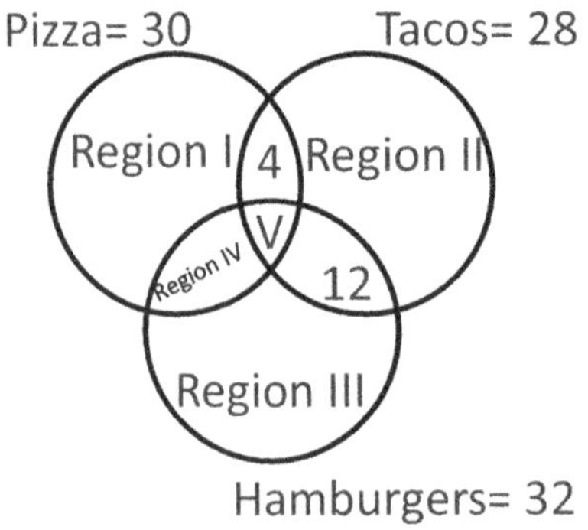

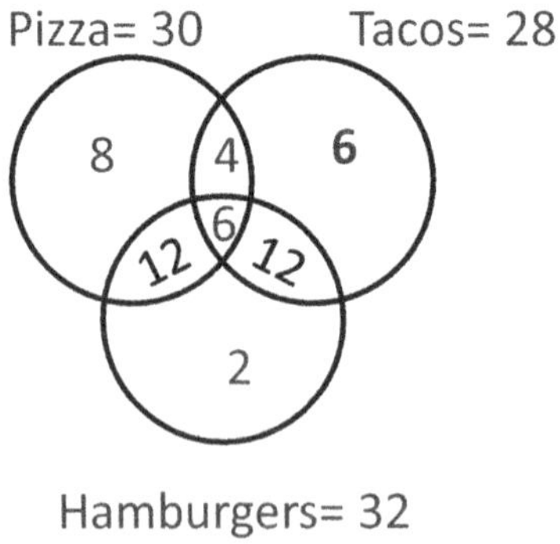

6

Name________________________________

PERENNIAL MATH

TEST #4

April 2019 Intermediate Level

1) If I triple the number N and divide the result by 9, the quotient is 20. What is N?	
2) A stake is driven in the ground so that 0.6 of it is sticking above ground. If eight inches is below ground, how many inches long is the entire stake?	
3) Three consecutive numbers add up to 174. What is the smallest of the three numbers?	
4) In the pattern below, the digits 1-9 are used and the pattern repeats after the last of the nine 9s. What will be the 100th digit if the pattern were written out? 1223334444......	
5) My seven friends line up. I write B for each boy and G for each girl. I have 3 B's and 4 G's, with no 2 B's in a row. In how many different orders can the B's and G's be written?	
6) Jeff has six more 40 cent stamps than 30 cent stamps. The total value of his 40 cent stamps is \$3.60 more than his 30 cent stamps. How many 40 cent stamps does Jeff have?	

PERENNIAL MATH

TEST #4

April 2019 Intermediate Level Answers

1) If I triple the number *N* and divide the result by 9, the quotient is 20. What is *N*?	60
2) A stake is driven in the ground so that 0.6 of it is sticking above ground. If eight inches is below ground, how many inches long is the entire stake?	20
3) Three consecutive numbers add up to 174. What is the smallest of the three numbers?	57
4) In the pattern below, the digits 1-9 are used and the pattern repeats after the last of the nine 9s. What will be the 100th digit if the pattern were written out? 1223334444……	4
5) My seven friends line up. I write B for each boy and G for each girl. I have 3 B's and 4 G's, with no 2 B's in a row. In how many different orders can the B's and G's be written?	10
6) Jeff has six more 40 cent stamps than 30 cent stamps. The total value of his 40 cent stamps is $3.60 more than his 30 cent stamps. How many 40 cent stamps does Jeff have?	18

PERENNIAL MATH

TEST #4

April 2019 Intermediate Level Solutions

1) If I triple the number N and divide the result by 9, the quotient is 20. What is N?

$$3N/9 = 20$$
$$3N/9 \times 9 = 20 \times 9$$
$$3N = 180$$
$$3N/3 = 180/3$$
$$N = 60$$

60

2) A stake is driven in the ground so that 0.6 of it is sticking above ground. If eight inches is below ground, how many inches long is the entire stake? Note that 0.6 is equivalent to 3/5 above ground. The 8 inches that are below ground is equal to 0.4 or 2/5 of the stake. If 2/5 is equal to 8 inches, then 4 inches is equivalent to 1/5 of the stake. There are 4 inches x 5 = 20 inches in the entire stake.

20

3) Three consecutive numbers add up to 174. Use algebra. The first number is X, the second number is X + 1, and the third number is X + 2. Add the numbers together and solve for X.

$$X + (X + 1) + (X + 2) = 174$$
$$3X + 3 = 174$$
$$3X = 171$$
$$X = 57$$

Since X represents the first number, then 57 is the answer.

57

PERENNIAL MATH

TEST # 4

April 2019 Intermediate Level Solutions

4) In the pattern below, the digits 1-9 are used and the pattern repeats after the last of the nine 9s. To find the 100th digit if the pattern were written out, see how many numbers are in the original pattern. The pattern is 1 + 2 + 3 + 4 + 5 + 6 + 7 + 8 + 9 = 45 numbers in the original pattern. So 100/45 numbers equals 2 with 10 numbers left over. The tenth remaining number when the pattern starts over is 4. So the 100th digit is 4.

1223334444......

4

5) Make a list of the possible line-ups starting with a boy (B) in the 1st position and then make a list with a girl (G) in the 1st position. The possibilities when lining up 3 B's and 4 G's, with no 2 B's in a row are:

1) BGBGBGG, 2) BGBGGBG, 3) BGBGGGB, 4) BGGBGBG, 5) BGGBGGB, 6) BGGGBGB,
7) GBGBGBG, 8) GBGBGGB, 9) GBGGBGB, and 10) GGBGBGB.

10

6) Jeff has six more 40¢ stamps than 30¢ stamps. The total value of his 40¢ stamps is $3.60 more than his 30¢ stamps. How many 40¢ stamps does Jeff have? Make a chart. Start with one 30¢ stamp and increase the number of 30¢ stamps until the difference in value between the two is $3.60. Jeff has 18 of the 40¢ stamps.

# of 30 cents	1	2	3	4	5	6	7	8	9	10	11	12
# of 40 cents	7	8	9	10	11	12	13	14	15	16	17	18
Value of 30 cent stamps	0.3	0.6	0.9	1.2	1.5	1.8	2.1	2.4	2.7	3	3.3	3.6
Value of 40 cent stamps	2.8	3.2	3.6	4	4.4	4.8	5.2	5.6	6	6.4	6.8	7.2
Difference in value	2.5	2.6	2.7	2.8	2.9	3	3.1	3.2	3.3	3.4	3.5	3.6

Note: The difference in value increases by 10¢ each time you increase the number of 30¢ stamps. $3.60 - $2.50 = $1.10 or 11 additional stamps. So, Jeff has 7 + 11 = 18 of the 40¢ stamps.

18

Name________________________________

PERENNIAL MATH

TEST #1

November 2019 Intermediate Level

1) The factory produces 300 flags in 15 minutes. How many flags can it produce in 1 minute?

2) A $12\frac{1}{2}$ foot pipe is cut into 6 equal pieces. How long is each piece in inches?

3) There are 3 different roads from Anniston to Birmingham, and 4 different roads from Birmingham to Cullman. How many different routes could Isaac plan from Anniston to Cullman?

4) If the area of one of the small squares is 36 cm^2, what is the perimeter of the large rectangle?

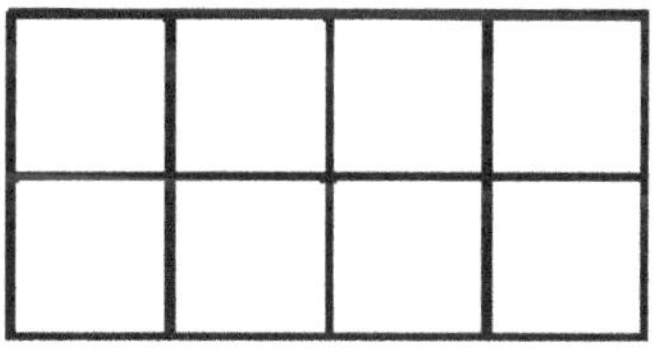

5) Jesse wants to have a least a 98 average in math. If she has earned a 100, 97, and 96 on the first three tests, what is the least score she can earn on the fourth test to achieve her goal?

6) Fred has 2 tanks. The smaller tank holds 336 gallons, but is only five-sixths full. If Fred empties this amount into the larger tank, it will only be four-ninths full. How many gallons will the larger tank hold when it is completely full?

PERENNIAL MATH

TEST #1

November 2019 Intermediate Level Answers

1) The factory produces 300 flags in 15 minutes. How many flags can it produce in 1 minute?	20
2) A $12\frac{1}{2}$ foot pipe is cut into 6 equal pieces. How long is each piece in inches?	25
3) There are 3 different roads from Anniston to Birmingham, and 4 different roads from Birmingham to Cullman. How many different routes could Isaac plan from Anniston to Cullman?	12
4) If the area of one of the small squares is 36 cm^2, what is the perimeter of the large rectangle? 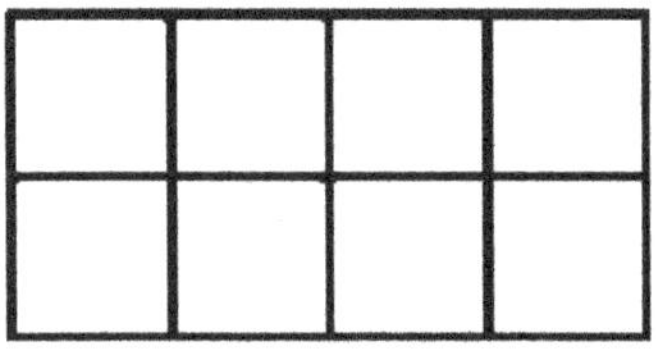	72
5) Jesse wants to have a least a 98 average in math. If she has earned a 100, 97, and 96 on the first three tests, what is the least score she can earn on the fourth test to achieve her goal?	99
6) Fred has 2 tanks. The smaller tank holds 336 gallons, but is only five-sixths full. If Fred empties this amount into the larger tank, it will only be four-ninths full. How many gallons will the larger tank hold when it is completely full?	630

PERENNIAL MATH

TEST # 1

November 2019 Intermediate Level Solutions

1) Flags per unit time = 300 flags / 15 minutes = 300/15 [flags per minute] = 20 flags per minute. Thus, the factory produces 20 flags in one minute.

20

2) A $12\frac{1}{2}$ foot pipe is cut into 6 equal pieces. How long is each piece in inches? First convert the $12\frac{1}{2}$ foot pipe into inches. 12 x 12 inches = 144 inches, plus the additional 6 inches in a half foot = 150 inches. Now divide the pipe into 6 pieces. 150 inches ÷ 6 = 25 inches. The 6 equal pieces are all 25 inches long.

25

3) There are 3 different roads from Anniston to Birmingham, and 4 different roads from Birmingham to Cullman. How many different routes could Isaac plan from Anniston to Cullman? Act this out. There are 12 routes.

The routes are:

AB1-BC1	AB2-BC1	AB3-BC1
AB1-BC2	AB2-BC2	AB3-BC2
AB1-BC3	AB2-BC3	AB3-BC3
AB1-BC4	AB2-BC4	AB3-BC4

12

PERENNIAL MATH TEST #1

November 2019 Intermediate Level Solutions

4) If the area of one of the small squares is 36 cm^2, what is the perimeter of the large rectangle? Since the area of one of the small squares is 36 cm^2 its sides are equal to 6cm. Therefore the length of the rectangle is equal to 4 x (6cm) = 24 cm. The width is equal to 2 x (6cm) = 12 cm. The perimeter of the rectangle is 2(24 cm +12 cm) = 72cm.

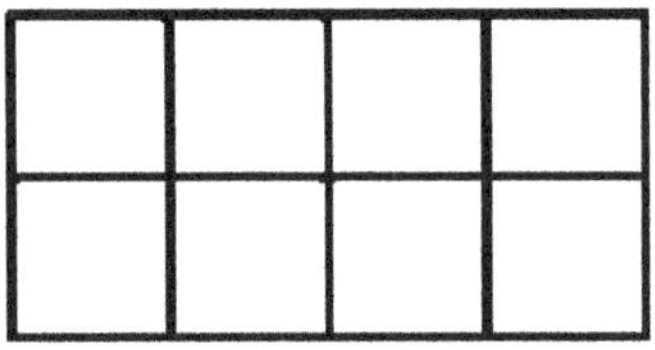

72

5) Jesse wants to have a least a 98 average in math. If she has earned a 100, 97, and 96 on the first three tests, she has earned 100 + 97 + 96 = 293 total points. To make a 98 average she must have 4 x 98 = 392 points. 392 – 293 = 99. The least score she can earn on the fourth test to achieve her goal is 99.

99

6) Fred has 2 tanks. The smaller tank holds 336 gallons, but is only five-sixths full. If Fred empties this amount into the larger tank, it will only be four-ninths full. How many gallons will the larger tank hold when it is completely full? First solve for how many gallons were in the smaller tank. 5/6 of 336 = 280 gallons. Now solve for the size of the larger tank. Since 280 gallons is 4/9, then 1/9 = 70 gallons. 70 x 9 = 630 gallons. The larger tank holds 630 tanks.

630

Name________________________________

PERENNIAL MATH

TEST #2

December 2019 Intermediate Level

1) What percentage of the letters in "DIVIDEND" are vowels? ____%	
2) What is the mean + median + mode + range of the following data set: 3, 4, 2, 6, 3, 3, 7	
3) How many triangles are in this shape? 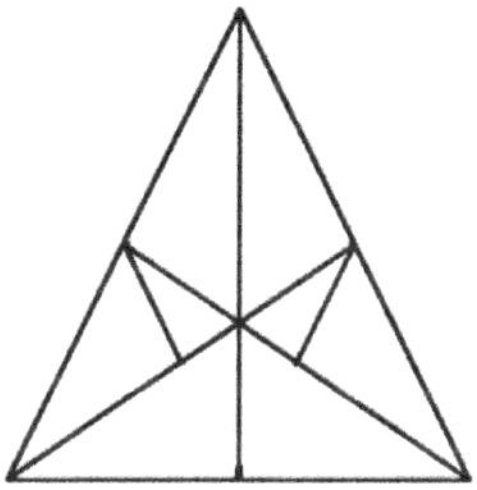	
4) Find the sum: 1 + 2 + 3 + 4 + ... + 87 + 88.	
5) In Mrs. Brown's class, 25 students like math and 17 students like science. Of all these students, 12 students like both math and science. How many total students are in the class?	
6) There are 6 types of pizza toppings at Giovanni's Pizza, but you may only select 3. How many combinations of three different toppings can you make?	

PERENNIAL MATH

TEST #2

December 2019 Intermediate Level Answers

1) What percentage of the letters in "DIVIDEND" are vowels? ____%	37.5
2) What is the mean + median + mode + range of the following data set: 3, 4, 2, 6, 3, 3, 7	15
3) How many triangles are in this shape? 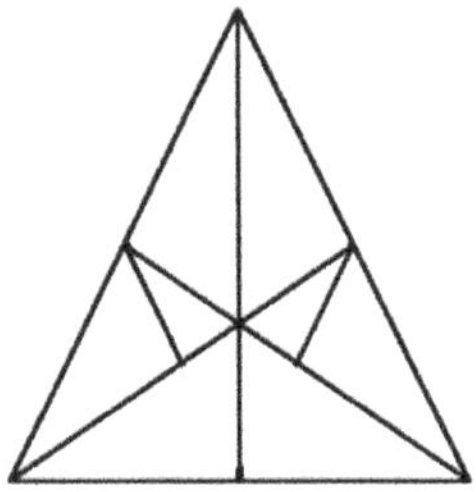	20
4) Find the sum: 1 + 2 + 3 + 4 + ... + 87 + 88.	3916
5) In Mrs. Brown's class, 25 students like math and 17 students like science. Of all these students, 12 students like both math and science. How many total students are in the class?	30
6) There are 6 types of pizza toppings at Giovanni's Pizza, but you may only select 3. How many combinations of three different toppings can you make?	20

PERENNIAL MATH TEST #2

December 2019 Intermediate Level Solutions

1) What percentage of the letters in "DIVIDEND" are vowels?
Three out of 8 of the letters in DIVIDEND are vowels.

3/8 = x/100 {cross multiple}
8x = 300
x = 37.5%.

37.5

2) Find the mean + median + mode + range of the following data set:
3, 4, 2, 6, 3, 3, 7.
First, put the data in order from lowest to highest 2, 3, 3, <u>3</u>, 4, 6, 7.
The mean is the average: (3 + 4 + 2 + 6 + 3 + 3 + 7) ÷ 7 = 4
The median is the middle number which happens to be 3.
The mode is 3 because it is the number that appears the most.
The range is difference of the highest and lowest numbers 7 – 2 = 5.
So, mean + median + mode + range = 4 + 3 + 3 + 5 = 15.

15

3) Count the shapes systematically. There are 8 individual triangles; 3 triangles made of two shapes; 2 triangles made of three shapes; 6 triangles made of four shapes; and 1 triangle made of all 8 shapes. There are 20 total triangles.

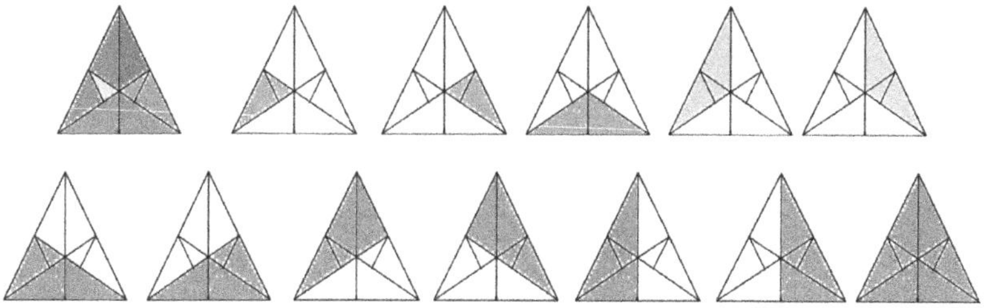

20

PERENNIAL MATH TEST #2

December 2019 Intermediate Level Solutions

4) Find the sum: 1 + 2 + 3 + 4 + ... + 87 + 88. You could add these up individually, or choose to do a short cut. If the numbers are paired from both ends there would be 44 pairs of numbers whose sum is 89. (1 + 88), (2 + 87), etc. Therefore, 44 x 89 = 3916.

3916

5) In Mrs. Brown's class, 25 students like math and 17 students like science. Of all these students, 12 students like both math and science. To find the number of students in the class use a Venn Diagram. First solve for the students who like Math only, 25 – 12 = 13. Next solve for the students who like Science only, 17 – 12 = 5. There are 13 + 12 + 5 = 30 students in the class.

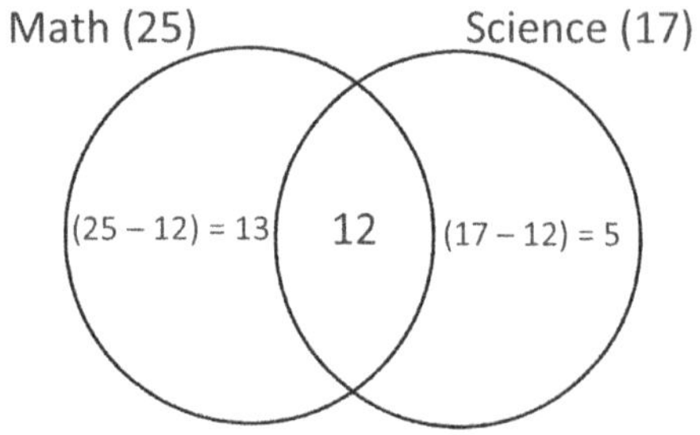

30

6) There are 6 types of pizza toppings at Giovanni's Pizza, but you may only select 3. How many combinations of three different toppings can you make? Make an organized list of the possible combinations of toppings 1-6. There are 20 possible three-topping pizzas.

123	234	345	456
124	235	346	
125	236	356	
126	245		
134	246		
135	256		
136			
145			
146			
156			

20

Name______________________________

PERENNIAL MATH

TEST #3

January 2020 Intermediate Level

1) 5 x 150 = 7500 ÷ ____.	
2) The digits 2, 4, 6, and 9 are arranged to form 2 two-digit numbers. Each digit may only be used once. If the 2 two-digit numbers are multiplied, what is the smallest product that can be obtained?	
3) In this cryptarithm, different letters represent different digits. What is the four-digit number ABCD? $\begin{array}{r} ABCD \\ BCD \\ CD \\ +\quad D \\ \hline AAAA \end{array}$	
4) This season the Warriors alternately won and lost all of their 101 games. If they won 13 of their first 25 games, then they won ___ of their last 15 games?	
5) On the way to Arizona, Jimmy traveled at 60 mph for 4 hours, 70 mph for 6 hours, and 57 mph for 2 hours. What was his average speed?	
6) Mrs. Brown gives each of her students 5 markers and there are 13 markers left over. However, if three students are absent, then each of the students would receive 6 markers with 10 markers left over. How many markers does Mrs. Brown have?	

PERENNIAL MATH

TEST #3

January 2020 Intermediate Level Answers

1) 5 x 150 = 7500 ÷ ____.	10
2) The digits 2, 4, 6, and 9 are arranged to form 2 two-digit numbers. Each digit may only be used once. If the 2 two-digit numbers are multiplied, what is the smallest product that can be obtained?	1,274
3) In this cryptarithm, different letters represent different digits. What is the greatest possible four-digit number ABCD? $\begin{array}{r} ABCD \\ BCD \\ CD \\ +\quad D \\ \hline AAAA \end{array}$	6319 or 8427
4) This season the Warriors alternately won and lost all of their 101 games. If they won 13 of their first 25 games, then they won ___ of their last 15 games?	8
5) On the way to Arizona, Jimmy traveled at 60 mph for 4 hours, 70 mph for 6 hours, and 57 mph for 2 hours. What was his average speed?	64.5
6) Mrs. Brown gives each of her students 5 markers and there are 13 markers left over. However, if three students are absent, then each of the students would receive 6 markers with 10 markers left over. How many markers does Mrs. Brown have?	118

PERENNIAL MATH

TEST # 3

January 2020 Intermediate Level Solutions

1) 5 x 150 = 7500 ÷ ____.

5 x 150 = 750. 7500 ÷ 750

5 x 150 = 7500 ÷ 10

10

2) The digits 2, 4, 6, and 9 are arranged to form 2 two-digit numbers. Each digit may only be used once. If the 2 two-digit numbers are multiplied, what is the smallest product that can be obtained? First, make the smallest numbers possible: 26 and 49, or 29 and 46. The product of 26 x 49 = 1,274. The product of 29 x 46 = 1,334. Therefore, the smallest product that can be obtained is 1,274.

1,274

3) Use trial and error to solve for the four-digit number ABCD. The letter A cannot be zero and it must be even since 4D = A. D cannot be 0 or 5 because A cannot be 0. Make a chart that starts with the possible values for D and whether or not the values will work for the other letters. There are two possible solutions. ABCD can equal 8427 or 6319.

```
  A B C D
    B C D
      C D
+       D
---------
  A A A A
```

D	A	C	B	Does it Work
1	4	8	1 or 6	No
2	8	6		No
3	2	7	5	No
4	6	5		No
6	4	4		No
7	8	2	4	Yes
8	2	3		No
9	6	1	3	Yes

8427 or 6319

PERENNIAL MATH TEST #3

January 2020 Intermediate Level Solutions

4) This season the Warriors alternately won and lost all of their 101 games. If they won 13 of their first 25 games, then they won ___ of their last 15 games? Since the Warriors won 13 of their first 25 games, they won the first game and every odd game since 13 + 12 = 25. Therefore, they won the odd games of the last 15 games which would be 8 of the 15 games. (game 101, 99, 97, 95, 93, 91, 89, 87).

8

5) On the way to Arizona, Jimmy traveled at 60 mph for 4 hours, 70 mph for 6 hours, and 57 mph for 2 hours. To find his average speed first solve for how far he went and then divide by how long it took him. He travelled (60 x 4 = 240 miles) + (70 x 6 = 420 miles) + (57 x 2 = 114 miles). 240 + 420 + 114 =774 miles. It took him 4 hours + 6 hours + 2 hours = 12 hours. 774miles/12 hours = 64.5 mph.

64.5

6) Mrs. Brown gives each of her students 5 markers and there are 13 markers left over. However, if three students are absent, then each of the students would receive 6 markers with 10 markers left over. How many markers does Mrs. Brown have?
Use algebra. Let S represent the number of students present and S – 3 represents the number of students when three students are absent. Write an equation with both sides equal since the number of markers does not change:

$$6(S - 3) + 10 = 5S + 13$$
$$6S - 18 + 10 = 5S + 13$$
$$6S - 8 = 5S + 13$$
$$S = 21$$

Mrs. Brown started with 5 x 21 +13 = 118 markers.

118

Name________________________________

PERENNIAL MATH

TEST #4

February 2020 Intermediate Level

1) How many two-digit numbers are divisible by 3?

2) The area of the figure below is 256 cm^2. What is the perimeter of one of the smaller squares?

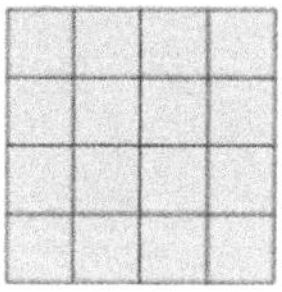

3) A 28oz bottle of canola oil is $5.88. If a 12oz bottle of canola oil has the same unit price as the larger bottle, how much would it cost?

4) On Saturday night, half the students in Mrs. Conley's class went to the dance, 1/3 of the rest of the students went to the movies, and the remaining 6 students stayed home. How many students are in Mrs. Conley's class?

5) Eric bought a new watch with a 20% off coupon. Its original price was $60 and local tax of 8% was charged after the 20% coupon was applied. What was the final price that Eric paid for the watch?

6) The sum of 1 + 3 + 5 + ... + 25 + 27 + 29 = 225.
Find the sum of 1 + 6 + 11 ... + 71 + 76 + 81.

PERENNIAL MATH

TEST #4

February 2020 Intermediate Level Answers

1) How many two-digit numbers are divisible by 3?	30
2) The area of the figure below is 256 cm^2. What is the perimeter of one of the smaller squares? 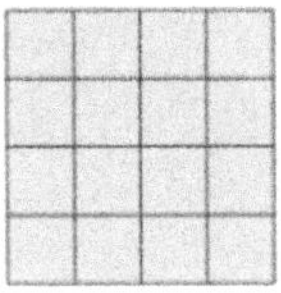	16
3) A 28oz bottle of canola oil is $5.88. If a 12oz bottle of canola oil has the same unit price as the larger bottle, how much would it cost?	2.52
4) On Saturday night, half the students in Mrs. Conley's class went to the dance, 1/3 of the rest of the students went to the movies, and the remaining 6 students stayed home. How many students are in Mrs. Conley's class?	18
5) Eric bought a new watch with a 20% off coupon. Its original price was $60 and local tax of 8% was charged after the 20% coupon was applied. What was the final price that Eric paid for the watch?	51.84
6) The sum of 1 + 3 + 5 + ... + 25 + 27 + 29 = 225. Find the sum of 1 + 6 + 11 ... + 71 + 76 + 81.	697

PERENNIAL MATH

TEST # 4

February 2020 Intermediate Level Solutions

1) How many two-digit numbers are divisible by 3? Numbers are divisible by 3 if the sum of all the individual digits is evenly divisible by 3. The numbers include 12, 15, 18, 21, 24, 27, 30, 33, 36, 39, 42, 45, 48, 51, 54, 57, 60, 63, 66, 69, 72, 75, 78, 81, 84, 87, 90, 93, 96, and 99. There are 30 numbers.

30

2) The area of the figure below is 256 cm^2. What is the perimeter of one of the smaller squares?
First solve for the dimensions of the figure. If the area of the figure below is 256 cm^2, then the sides equal 16 since it is the square root of 256. If the sides equal 16, then the length can be divided by 4 to find the length of the side of one of the smaller squares. 16 cm/4 = 4 cm. The perimeter of one of the smaller squares is 4 cm x 4 = 16 cm.

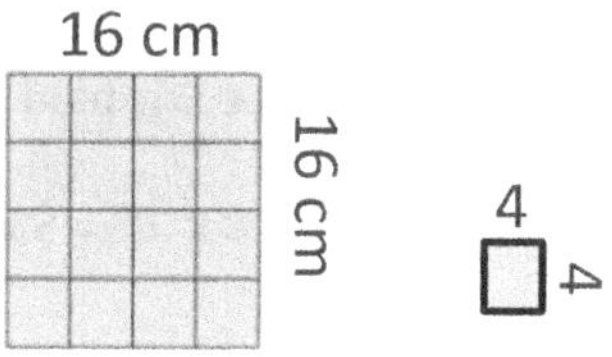

16

3) A 28oz bottle of canola oil is $5.88. If a 12oz bottle of canola oil has the same unit price as the larger bottle, how much would it cost? First divide to find the price per ounce. $5.88/28 = .21 per ounce. Now multiply to find the cost of the 12oz bottle. .21 x 12 = $2.52.

2.52

PERENNIAL MATH

TEST #4

February 2020 Intermediate Level Solutions

4) On Saturday night, half the students in Mrs. Conley's class went to the dance, 1/3 of the rest of the students went to the movies, and the remaining 6 students stayed home. How many students are in Mrs. Conley's class? Work backwards. Working only with the ½ of the students who did not go to the dance. 1/3 of them went to the movie and then there were 6 students remaining. 6 students is 2/3 of that half. So there are 9 students in that half. 9 x 2 = 18 students.

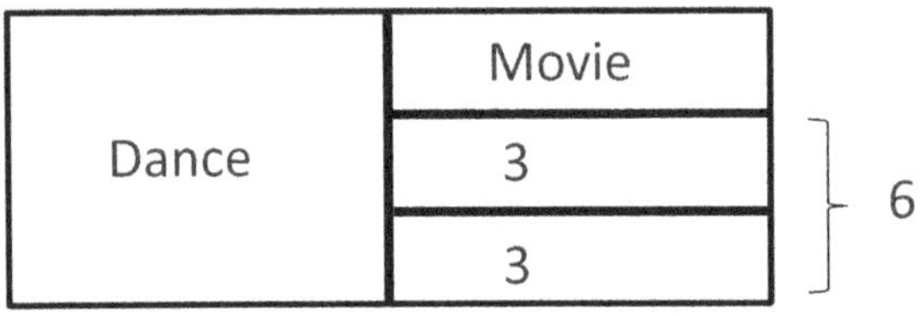

18

5) Eric bought a new watch with a 20% off coupon. Its original price was $60 and local tax of 8% was charged after the 20% coupon was applied. What was the final price that Eric paid for the watch? First find the price Eric paid for the watch after the discount of 20%. $60 x .20 = $12.
$60 - $12 = $48. Now compute the 8% tax. $48 x .08 = $3.84. $48 + $3.84 = $51.84.

51.84

6) The sum of 1 + 3 + 5 + ... + 25 + 27 + 29 = 225.
Find the sum of 1 + 6 + 11 ... + 71 + 76 + 81.
In the series 1 + 6 + 11 ... + 71 + 76 + 81, we get from 1 to 81 by adding 5 fourteen times, so the series has 15 terms. If you pair beginning and ending terms you will find eight pairs that add to 77 and then the last number 81: (1 + 76), (6 + 71), (11 + 66), (16 + 61), (21 + 56), (26 + 51), (31 + 46), (36 + 41), + 81 = 697.

697

Made in the USA
Monee, IL
05 November 2026

67811038R00036